Elementary Physics

Solids

San Diego • Detroit • New York • San Francisco • Cleveland • New Haven, Conn. • Waterville, Maine • London • Munich

For more information, contact
The Gale Group, Inc.
27500 Drake Rd.
Farmington Hills, MI 48331-3535
Or you can visit our Internet site at http://www.gale.com

Photo Credits: **Art I Need:** 1, 4; **The Brown Reference Group plc:** 2t, 8, 10, 16; **Corbis:** Rick Gayle 14; **Hemera Photo Objects:** 2c, 2b, 3t, 6, 7; **Photodisc:** C. Borland/Photolink 3b, Robert Glusic 12, 18.

Consultant: Don Franceschetti, Ph.D., Distinguished Service Professor, Departments of Physics and Chemistry, The University of Memphis, Memphis, Tennessee

For The Brown Reference Group plc
Text: Ben Morgan
Project Editor: Tim Harris
Picture Researcher: Helen Simm
Illustrations: Darren Awuah and Mark Walker
Designer: Alison Gardner
Design Manager: Jeni Child
Managing Editor: Bridget Giles
Production Director: Alastair Gourlay
Children's Publisher: Anne O'Daly
Editorial Director: Lindsey Lowe

LIBRARY OF CONGRESS CATALOGING-IN-PUBLICATION DATA

Morgan, Ben.
Solids / by Ben Morgan.
p. cm. — (Elementary physics)
Includes bibliographical references and index.
ISBN 1-41030-085-4 (hardback: alk. paper) — ISBN 1-41030-203-2 (paperback: alk. paper)
1. Solids—Juvenile literature. 2. Plastics—Juvenile literature. [1. Solids.] I. Title.

QC176.3.M67 2003
530.4'1—dc21 2003002547

Printed and bound in Singapore
10 9 8 7 6 5 4 3 2 1

Contents

Gold bars are solid.

What Are Solids?

Everything in the world is made of matter. There are three types of matter—**solids**, **liquids**, and **gases**. Solids are things that are hard and keep their own shape, like rocks and metals. Homes are made out of solid materials like bricks, wood, and glass. Most of the furniture and other things in our homes are solids.

Liquids are things that you can pour, like water. They have no shape of their own. Instead, they take the shape of whatever you put them in. Gases, like air, are usually **invisible**. They cannot be seen or held. They fill all the empty spaces around us.

All the things on this page are solids.

Different Solids

There are many kinds of **solids**. Some of the most common kinds are **plastic**, glass, and wood. Solids have different **properties**. This means that some solids are heavy, while others are light. Some solids are hard, and others are soft.

Toys made of plastic are hard to break.

Metals are solids that are hard, heavy, and shiny. Glass is hard and heavy, too. It is also see-through. Glass is **brittle**, which means it can break easily. Plastic and wood are lighter than metal or glass. Plastic is **flexible**, which means it bends. Rubber bends even more than plastic.

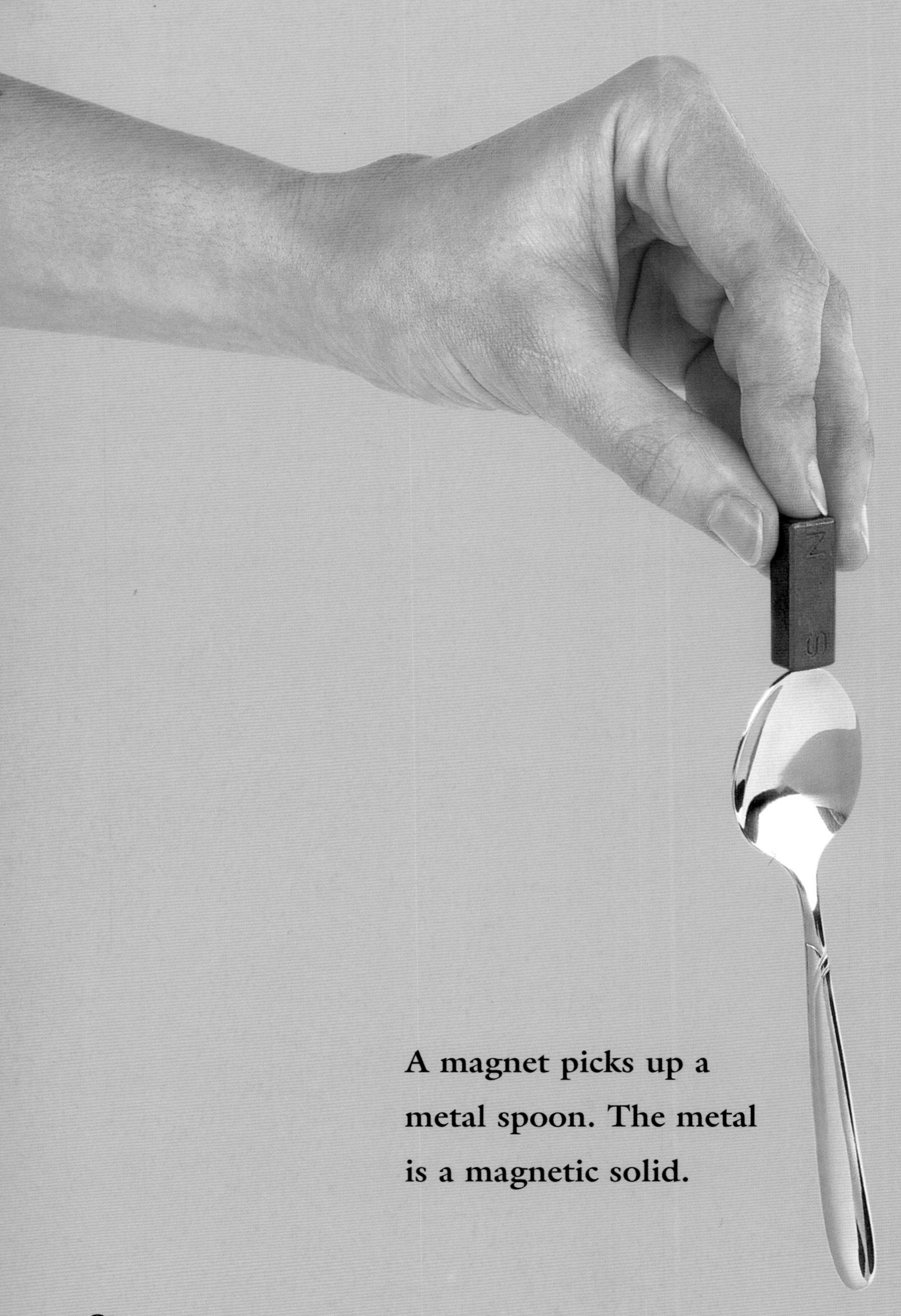

A magnet picks up a metal spoon. The metal is a magnetic solid.

Investigating Solids

It is easy to learn the **properties** of some **solids**. To find out that rocks are heavy, you only have to pick up a rock. Some properties are not so easy to learn, though. You can only discover these properties by carrying out **experiments**.

Certain solids are **magnetic**, for example. That means that a magnet can pull them, even without touching them. If you have a magnet, try to find out which types of solids are magnetic. Other solids float in water, while some sink. If you drop a handful of different solids in a sinkful of water, you will see which ones float.

These frozen treats are solids. They will melt if they get warmer.

Melting and Freezing

When **liquids** get very cold, they **freeze**. That means they turn into **solids**. Likewise, when solids get hot, they turn into liquids. They **melt**.

Some solids need to get hotter than others to melt. Chocolate melts in the warmth of your mouth, but a candle needs the heat of a flame to make its wax melt. Metal and glass melt in the heat of a powerful fire. As soon as they come out, they cool and turn back into solids.

Without strong metal cables, this bridge would not stand up.

Metals

Metals are very hard and strong. People use them to make things that have to last. Bridges, ships, cars, and coins are all made of metals. Although metals are hard, they are also slightly **flexible**. They can be pulled into wires. They can be rolled into thin sheets like aluminum foil.

Metals heat up quickly, so ovens and cooking pans are usually made of metal. Most metals are shiny. Shiny metals like gold and silver are often used to make jewelry.

Diamonds are
very hard crystals.

Crystals

Crystals are another kind of **solid**. They have flat surfaces and sharp edges and corners. Diamonds and rubies are crystals. So are sand grains and snowflakes. If you look closely at a snowflake, you can see that it is made of tiny crystals joined in a ring. Rocks and pebbles are made of lots of small crystals packed together. If you look closely at a pebble, you might see the crystals glinting like jewels.

Diamonds are the hardest crystals. Crystals are also brittle. They can split and shatter. Salt and sugar are crystals, too. If you stir salt or sugar in water, they **dissolve,** or disappear into the water.

All the things on this
page are flexible solids.

Plastics

Plastics are light, **flexible solids**. They are lighter than **metals** and less brittle than crystals. But they are still very strong. Plastics are easy to color and mold into different shapes. That is why plastics are used to make all sorts of things, from plastic bags and clothes to toys and fridges.

Plastic things are usually slightly flexible. Gently squeeze or bend a plastic object. You will find it springs back into shape afterward. Be careful though—it might snap if you bend it too far! Plastics are not natural, like wood or rock. People make them by mixing chemicals together.

This giant tree contains air and water, so it is really a mixture of solids, gases, and liquids.

Living Solids

Wood and bone are natural **solids**. They form inside living things, including people. The bones inside your body make up your **skeleton**. The skeleton supports your body and gives you shape. Without a skeleton, your body would be like a blob of wobbly gelatin. Wood does a similar job. The trunk and branches of a tree hold the leaves high in the air. Without wood, a tree would be a heap of leaves.

Natural solids are mixtures of different materials. Living bone, for example, contains blood, so it is a mixture of solids and **liquids**.

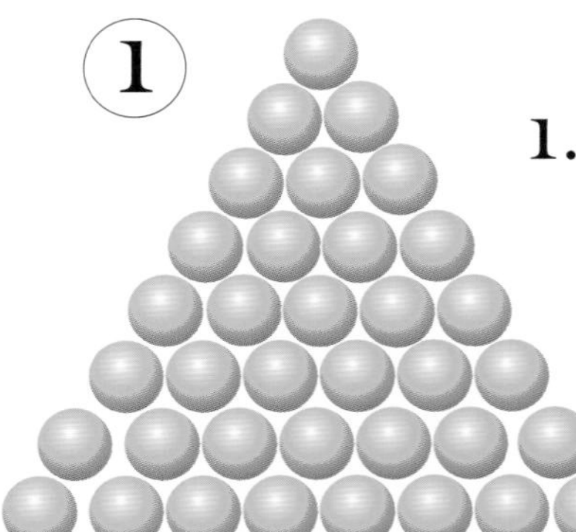

1. The molecules in a solid are packed together like bricks in a wall.

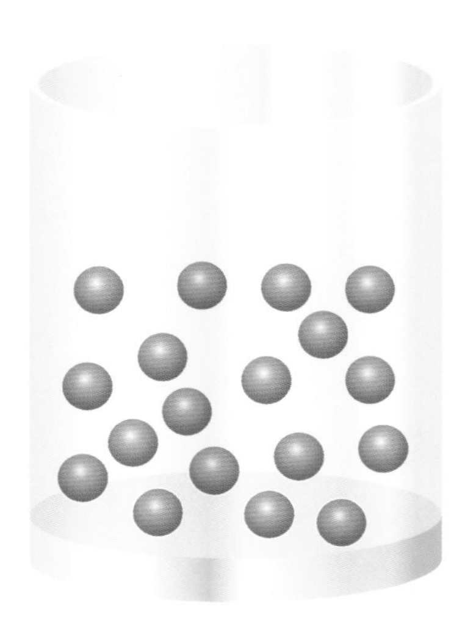

2. The molecules in a liquid can move around separately, but they always stay close together.

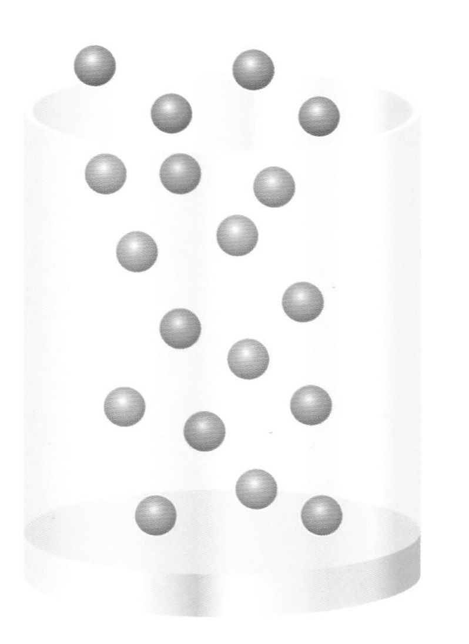

3. The molecules in a gas fly off on their own and spread out.

Atoms and Molecules

Everything in the world is made of very tiny particles called **atoms**. Usually the atoms form small groups called **molecules**. Molecules are so small that you cannot see them, even with a **microscope**. **Solids**, **liquids**, and **gases** are different from each other because of the way their molecules are arranged. In a solid, all the molecules are joined together strongly and held in place like bricks in a wall. In a liquid, the molecules can move around, but they always stay close together. In a gas, the molecules can fly off on their own and spread out.

Bouncing Experiment

Carry out this **experiment** to test the **properties** of different solids. Collect a range of different things, such as balls, plastic toys, a rubber band, a lump of modeling clay, and a sock. Do not use anything that is valuable or made of glass. Hold each item as high as you can with your arm. Then let go. Which things bounce the highest?

Round and flexible things bounce the best. When they hit the ground, they get squashed a bit. This pushes them back up. Things that are too brittle to get squashed hardly bounce. Things that are too soft to spring back into shape will not bounce either.

Glossary

atom the smallest particle of a liquid, solid, or gas. Atoms can group together to form molecules.

brittle easily snapped or cracked.

dissolve to disappear into a liquid.

experiment (noun) a test or trial.

experiment (verb) to test or try out.

flexible easily bendable.

freeze turn from a liquid into a solid.

gas one of the three states of matter. Gases have no shape and spread out to fill space.

invisible something that is there but cannot be seen.

liquid one of the three states of matter. A liquid can be poured into a container.

magnetic something that can be pulled by a magnet.

melt turn from a solid into a liquid.

metal a hard, shiny, usually solid material, such as iron or gold.

microscope an instrument to look at very small objects.

molecule a group of atoms.

plastic a light, flexible solid.

property the nature of a gas, liquid, or solid.

skeleton the bones in the body.

solid one of the three states of matter. Solids are hard and keep their shape.

Look Further

To find out more experiments you can carry out with solids, read *101 Great Science Experiments* by Neil Ardley (DK Publishing).

You can also find out more about the states of matter from the internet at this website: www.chem4kids.com/

Index

DATE DUE

DEC 14 2006			
DEC 15			